KV-575-363

WORLD ABOUT US
FOSSIL FUELS

MARGARET SPENCE

BRIDGE PRIMARY SCHOOL
BANBRIDGE

LIBRARIES NI
WITHDRAWN FROM STOCK

GLOUCESTER PRESS
London·New York·Toronto·Sydney

© Aladdin Books Ltd 1993

Designed and produced by
Aladdin Books Ltd
28 Percy Street
London W1P 9FF

First published in
Great Britain in 1993 by
Watts Books
96 Leonard Street
London EC2A 4RH

Design: David West
Children's
Book Design
Designer: Stephen
Woosnam-
Savage
Editor: Fiona Robertson
Consultant: Brian Price,
Pollution
Consultant
Illustrator: Mike Lacy

ISBN 0 7496 0914 1

A CIP catalogue record for this
book is available from the
British Library.

Printed in Belgium
All rights reserved

Contents

Fossil fuels
4
What is coal?
6
Down the mineshaft
8
Oil and gas
10
Drilling for oil
12
Transporting
14
Generating electricity
16
Other uses
18
Disaster!
20
Waste not...
22
A cleaner generation
24
Into the future
26
Fact file
28
Glossary
31
Index
32

Introduction

Fossil fuels are an essential part of our lives today. However, for many years now, we have known that supplies of fossil fuels are limited. If we continue to use them at the present rate, the reserves that took millions of years to form will quickly run out. We cannot afford to let this happen. Remaining reserves of fossil fuels must be protected.

Fossil fuels

Fossil fuels have been described as the Earth's buried treasure. Formed millions of years ago from the remains of plants and animals that lived on Earth, coal, oil and gas are our most valuable sources of energy today. When we burn fossil fuels in power stations, we are changing the sunshine stored by plants and animals as they grew, into heat. This heat can then be converted into electricity.

When coal is heated without oxygen, coke is produced. Coke is a vital part of steel-making (see page 18).

Without oil, there would be no petrol or diesel on which to run our cars.

When fossil fuels are burned, they create polluting gases which add to the problems of acid rain and global warming (see page 21).

Fossil fuels provide us with energy to heat our homes and cook our food. Many industrial goods could not be made without the energy from fossil fuels.

Fossil fuels not only help to run our vehicles; they are also used in many of the materials that cars are made from (see page 23).

Gas and oil are used to make the pesticides and fertilisers (see page 19) which are needed to grow food to feed the world's increasing population.

5

What is coal?

Over 300 million years ago, much of the Earth was covered with thick swamps, dense forests and shallow seas. In the hot, humid climate, plants grew, died and decayed, forming a thick layer of rotting plant material at the bottom of the swamps. The plants gradually became buried under layers of silt and mud. Heat and pressure from these layers changed the plant material into coal.

Coal is found in layers called seams. Coal seams can be broken or their position altered, by earthquakes or other earth movements.

Peat is the first stage of coal formation. Peat can form just a few hundred years after the dead plants have been buried.

6

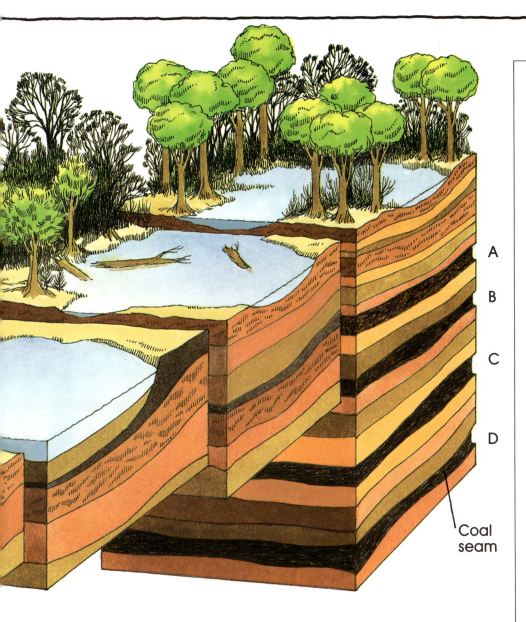

A
B
C
D

Coal seam

Types of coal

A. Lignite is formed from peat. It usually occurs near the surface.

B. Cannel is a harder, more brittle type of coal.

C. Bituminous coal is mostly used in homes and factories.

D. Anthracite is the hardest and most expensive type of coal.

The era when coal was formed is known as the Carboniferous Era.

It takes several metres of plant material to form a seam of bituminous coal that is just 30 cm thick.

Down the mineshaft

The best types of coal are usually found deep below the Earth's surface. Vertical tunnels called shafts are sunk down to the seams. For centuries, miners worked the coal face with a pick and shovel. Today, drills, cutting machines and computers do most of the work. Yet the world of the mine is still a dark and dangerous one. All miners wear protective clothing and a hard helmet with a lamp.

At the surface, huge fans force fresh air down one shaft and drive stale air up another. However, the threat of a gas leak leading to an explosion is a very real one. Miners leave behind a numbered disk which is collected when they return to the surface. In the event of an accident, the disk can show if the miner is still down the mine.

Types of mining
Strip mining is used to get coal lying near the surface. Unlike open-cast mining (also done to remove coal near the surface), the damage done by strip mining is temporary. As one piece of land is dug up, the rock and soil are used to cover an area which has already been mined.

Shaft mines have to be sunk to remove coal at deeper levels.

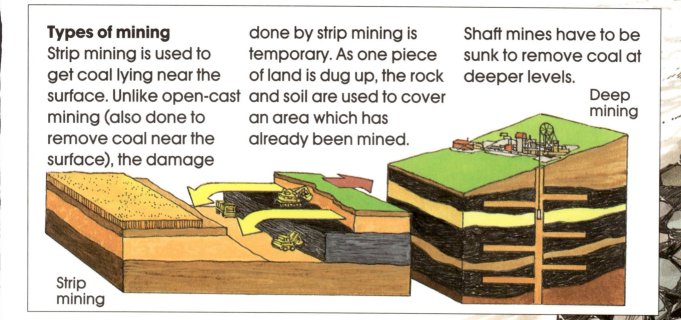

Strip mining

Deep mining

In the long wall method of mining (left), the roof is held up by supports, which are moved forward as the wall is mined.

The coal is taken by conveyor belt to the bottom of the shaft. It is then lifted to the surface in shuttle cars, called skips. From here, it may either be transported to a nearby power station, to industry, or to other countries.

Oil and gas

Oil and gas are hydrocarbons (a mixture of hydrogen and carbon). Like coal, they can be burned to produce electricity, although they create less pollution. Gas was first used as a fuel by the Chinese over 2,000 years ago. But it was not until 300 years ago that it was widely recognised as a source of heat and light. Oil is considered so valuable in our society today, it has been nicknamed "black gold". Most of our vehicles depend on fuels made from oil.

Oil and gas were formed in a similar way to coal. Tiny prehistoric plants and animals, called plankton, died and settled on the sea bed. Under the heat and pressure of many layers of mud and rock, the plants and animals changed into oil and gas. Oil and gas are often found together.

Plankton

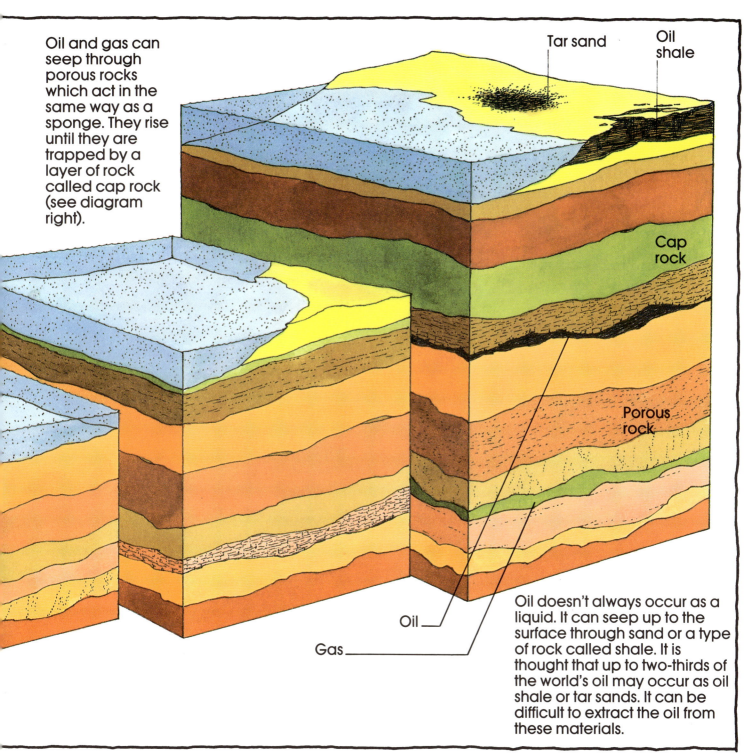

Oil and gas can seep through porous rocks which act in the same way as a sponge. They rise until they are trapped by a layer of rock called cap rock (see diagram right).

Tar sand

Oil shale

Cap rock

Porous rock

Oil

Gas

Oil doesn't always occur as a liquid. It can seep up to the surface through sand or a type of rock called shale. It is thought that up to two-thirds of the world's oil may occur as oil shale or tar sands. It can be difficult to extract the oil from these materials.

Drilling for oil

The search for oil is dangerous and expensive, especially offshore. Instruments called seismometers (below) test for oil in the layers of rock underground. If the results are good, a test well is drilled from an exploratory rig (right). The sides of the well are lined with steel to stop them collapsing. A type of chemical mud is pumped down the drill and forces tiny pieces of rock, along with any traces of oil, to the surface.

Most of the test wells at sea are towed from place to place by small boats. Once in position, the rigs are either secured by chains or fastened to the sea bed on long, spindly legs.

One ship sets off an explosion and another tows a cable carrying listening devices called hydrophones.

The shock waves from the explosion travel downwards to the Earth's crust. The shock waves bounce off the crust and are recorded onto a graph. The pattern (right) can show if oil is present.

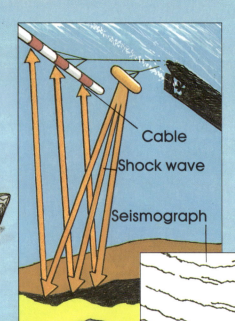

Cable

Shock wave

Seismograph

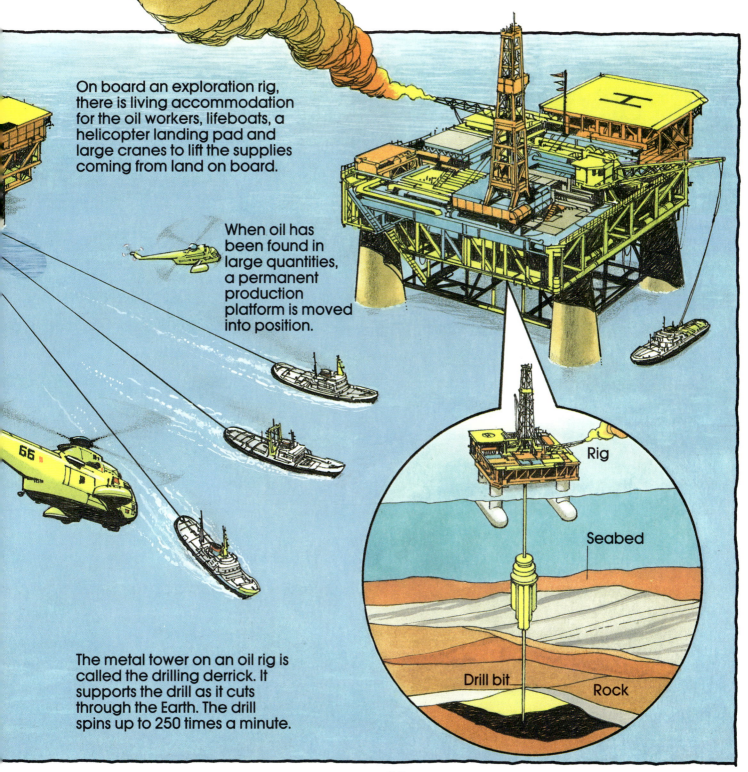

On board an exploration rig, there is living accommodation for the oil workers, lifeboats, a helicopter landing pad and large cranes to lift the supplies coming from land on board.

When oil has been found in large quantities, a permanent production platform is moved into position.

The metal tower on an oil rig is called the drilling derrick. It supports the drill as it cuts through the Earth. The drill spins up to 250 times a minute.

Rig

Seabed

Drill bit

Rock

Transporting

Every day, millions of barrels of crude oil are transported from the production platform to the refinery (see below). The most common ways of transporting oil are by pipelines on land or at sea, or by tankers at sea. The pipelines are usually made of steel and built to withstand great pressure. Because they are more expensive than tankers, pipelines tend to be used for transporting the oil over short distances.

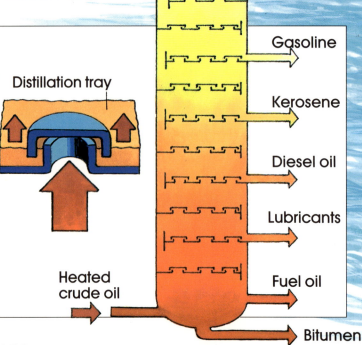

Gases

Petroleum

Gasoline

Kerosene

Diesel oil

Lubricants

Fuel oil

Bitumen

At the refinery

The different substances in crude oil are separated out by a process called "fractional distillation". The oil is boiled until it turns into a gas, and passed into a tall column. As the gas cools, the substances in it turn to liquid at different temperatures, and are drawn off.

Distillation tray

Heated crude oil

Natural gas often surfaces with oil. The two are separated and the gas is sent to a processing factory, where it is cleaned before it can be used.

Some of the oil tankers used are up to 400 m long and can carry nearly 500,000 tonnes of oil. The oil is stored in huge tanks which are filled with water when the oil is unloaded to keep the ship afloat.

Generating electricity

A major use for fossil fuels, particularly coal, is to generate electricity. Most power stations work in the same way. The fuel is burned to release heat, which is used to turn water into steam. The steam is then used to turn the blades of a massive rotating turbine. The turbine powers a generator which produces electricity. In many countries, the electricity is sent to the national grid.

The huge clouds of steam that can often be seen belching from power station towers are waste heat. Steam passing through the turbine is turned back into water in condensers.

The national grid system consists of a network of power stations, which are all linked. In this way power can be supplied where and when it is needed, and extra power stations can even be started up to meet peak demands.

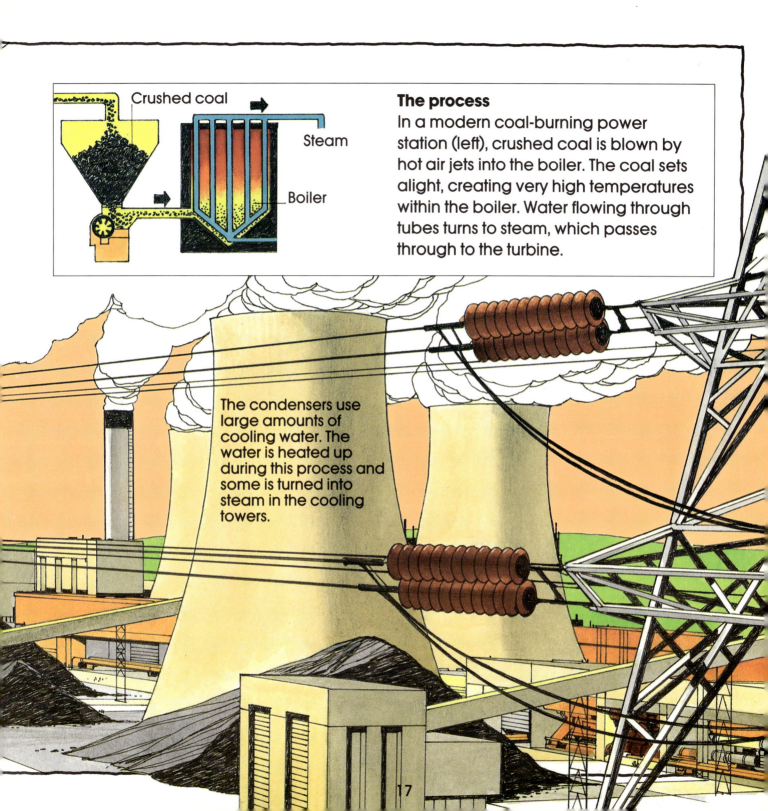

Crushed coal

Steam

Boiler

The process

In a modern coal-burning power station (left), crushed coal is blown by hot air jets into the boiler. The coal sets alight, creating very high temperatures within the boiler. Water flowing through tubes turns to steam, which passes through to the turbine.

The condensers use large amounts of cooling water. The water is heated up during this process and some is turned into steam in the cooling towers.

Other uses

The energy produced by fossil fuels is just one of the reasons they are considered so important. Coal can be heated without oxygen to produce coke, which is vital for making iron and steel. This process also creates the coal-tar pitch used in road-making. In hot, dry places, the heat from burning gas is used to extract the salt from sea-water. The water can then be used for drinking.

Fossil fuels turn up in the most unlikely places. Nylon clothes, plastic toys, detergents and even carpets are all made from substances that come from crude oil.

Some types of printing ink and glues are made from coal, and coal-tar soap is made from the coal-tar produced when coal is heated to make coke.

The chemicals produced when crude oil is refined are called petrochemicals. These chemicals are used widely in industry, from making drugs, to pesticides and animal food, to plastics.

Outside the home, the paint used on doors and walls comes from oil, and petrol or diesel is used in cars. The ash from burning coal can be made into breeze blocks, which may be used to build houses.

Disaster!

No fuel is perfect, and fossil fuels can create problems. Underground coal mining is one of the most hazardous jobs in the world. Roof falls, build-ups of gas (especially methane), explosions and flooding are still major threats, despite recent improvements. On land, the ground above old mines can begin to sink into the gap left below, making nearby houses and buildings unsafe. Every year, tanker accidents and leaking pipelines cause often devastating damage to the worlds's oceans and coastlines.

A fire, or "blow-out", at an oil rig can burn for years if it is not treated. Controlling a blow-out is a very dangerous job. The fire has to be blasted with explosives. The force of the explosion is intended to put out the fire.

In 1991, at the end of the Gulf war, retreating Iraqi soldiers placed dynamite in hundreds of Kuwaiti oil wells. More than 500 wells caught fire, causing severe damage to the nearby land and sea.

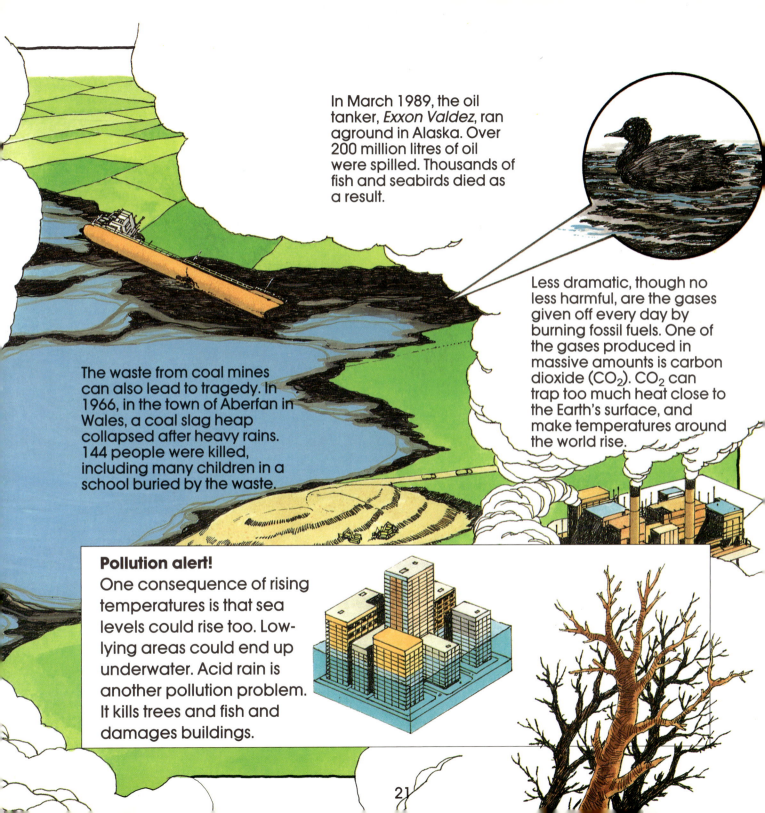

In March 1989, the oil tanker, *Exxon Valdez*, ran aground in Alaska. Over 200 million litres of oil were spilled. Thousands of fish and seabirds died as a result.

Less dramatic, though no less harmful, are the gases given off every day by burning fossil fuels. One of the gases produced in massive amounts is carbon dioxide (CO_2). CO_2 can trap too much heat close to the Earth's surface, and make temperatures around the world rise.

The waste from coal mines can also lead to tragedy. In 1966, in the town of Aberfan in Wales, a coal slag heap collapsed after heavy rains. 144 people were killed, including many children in a school buried by the waste.

Pollution alert!
One consequence of rising temperatures is that sea levels could rise too. Low-lying areas could end up underwater. Acid rain is another pollution problem. It kills trees and fish and damages buildings.

Waste not....

Recycling involves collecting the valuable materials in waste, and making them into new products. Because it saves energy and resources, recycling is a good solution to the problem of decreasing fossil fuel supplies. Glass, iron, aluminium and steel can all be recycled. Paper and other waste could be burnt in an incinerator to produce heat, also reducing the need for fossil fuels.

One of the keys to recycling is separating the different materials. Many towns now have separate bins for different kinds of rubbish.

If the 80 million aluminium cans which the world uses each year were replaced with refillable glass bottles, the amount of energy saved could provide electricity for 13 million people.

The metal for cars, the plastic for the seats, the dashboard, the gearstick and so on, are all made from precious reserves of fossil fuels. The German car maker, BMW, has now developed a car made completely from recyclable materials.

A cleaner generation

In the past 30 years, many improvements have made fossil fuels cleaner and more efficient. Research is being carried out to reduce the amount of sulphur dioxide and nitrogen oxide gases given off when coal is burned. One system, known as "flue gas desulphurisation", involves spraying the gases with special chemicals, which remove the sulphur in them. Cars can be fitted with catalytic converters which reduce harmful exhaust gases by a massive 95%.

If cars could be made just 20% more efficient, 200 million tonnes of oil would be saved each year. Cars can be specially designed to allow air to flow over them more easily. This is called stream-lining, and cuts down on the amount of fuel used.

Many manufacturers are working with the fossil fuel industry to develop cleaner and more efficient technology.

Today, most trains run on electricity or diesel. One possibility for the future is trains that run on steam produced by burning coal in a special boiler.

Trains and buses can carry more people more miles on the same amount of fuel than private cars.

Fluidised bed boilers

In a fluidised bed boiler, coal is burnt with sand (which contains lime-stone) in a chamber, through which air is blown. The limestone traps the sulphur that is produced, and stops it escaping. Harmful gases are reduced by 90%.

Crushed coal

Cold water

Natural gas out

Stirrer

Fluidised bed

Steam

Air

Hot gases

Steam and oxygen

Ash

Into the future

If we are to make the best use of the remaining supplies of fossil fuels, other sources of energy must be developed that can be used alongside coal, oil and gas. Wind, moving water and solar power are natural forms of energy. They can be used to produce electricity fairly cleanly, and will never run out. Geothermal energy, or energy that comes from the heat of the Earth's crust, can also be used as a source of power. We cannot live without energy. The fuel reserves remaining to us must therefore be carefully conserved.

As supplies of fossil fuels run low, schemes like the one below are being developed. This off-shore gas powered generating station uses gas that was previously considered too low quality to be profitable, to produce electricity.

The Sun's energy can be converted directly into electricity using rows of solar cells. Solar cells are already used in many parts of the world, and can be made to work even in cloudy weather.

Rows of floating rafts use the up-and-down movement of the waves to produce electricity. Computer-controlled sails are being installed on some ships to reduce the amount of fuel used.

Huge modern wind turbines have blades that are specially shaped to catch the wind's energy. However, it would take about 1,000 wind turbines to produce the same amount of energy as a medium-sized power station.

Fact file

The Trans-Alaska pipeline crosses the frozen lands of Alaska for nearly 1,300 km. The pipeline is raised on stilts to prevent the oil that is pumped through it, from melting the surrounding frozen ground. The stilts also mean that the pipeline does not block the path of migrating reindeer.

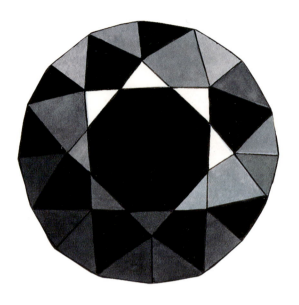

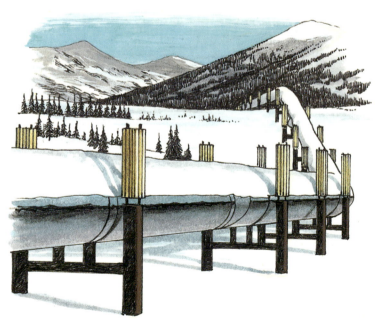

One of the more unlikely uses for coal is in jewellery. Jet is a type of bituminous coal which can be carved and polished to look like black glass. It is used to make buttons, costume jewellery, brooches and so on. During the 19th century, the town of Whitby in Yorkshire was famous for its production of jet. Today, jet can also be found in Spain, France and the US.

Over 50% of the world's known oil reserves are found in the Middle East. The former Soviet Union also has huge reserves of oil, as well as over 40% of the world's reserves of natural gas. Coal, on the other hand, is more widely distributed; it is found in every continent in the world. However, the countries with the greatest reserves are not always the largest producers of fossil fuels. Some countries may import fossil fuels to save their own reserves; others may be forced to stop producing coal, oil or gas because of economic problems.

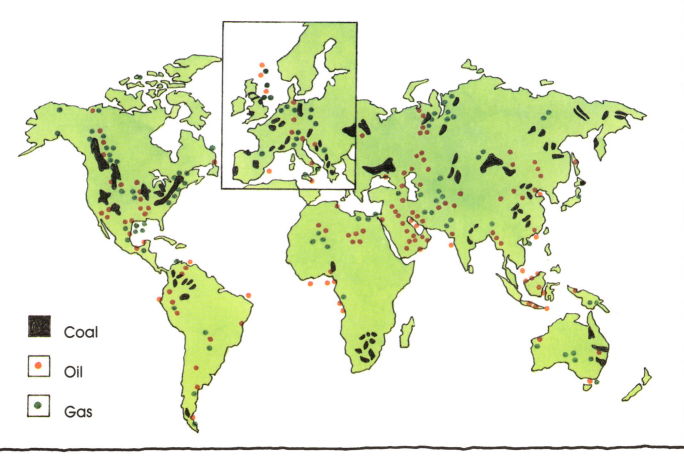

Coal

Oil

Gas

One of the systems developed to improve fossil fuel efficiency is known as the topping cycle. In the topping cycle, part of the coal is changed into a fuel gas, and burnt in a turbine. The remainder is burnt in a combustor to produce the steam needed for an ordinary generator. The result is that electricity is produced 20% more efficiently, and that the amount of harmful gases produced is greatly reduced.

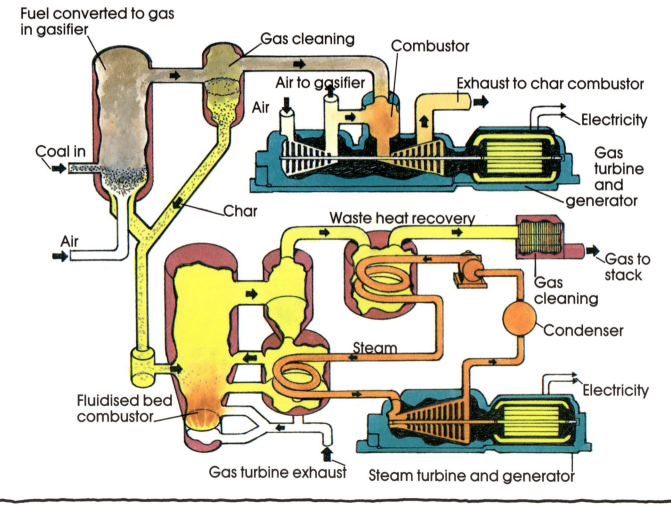

Fuel converted to gas in gasifier

Gas cleaning

Combustor

Air to gasifier

Air

Exhaust to char combustor

Electricity

Coal in

Char

Gas turbine and generator

Air

Waste heat recovery

Gas to stack

Gas cleaning

Condenser

Steam

Electricity

Fluidised bed combustor

Gas turbine exhaust

Steam turbine and generator

Glossary

Acid rain
The gases produced by burning fossil fuels can mix with the moisture in clouds to form a weak acid. This acid then falls to the ground with rain.

Coal tar
The thick, black liquid produced when coal is burned without oxygen.

Derrick
A metal tower found on an oil-drilling rig. It is placed over the drilling shaft.

Fossil fuels
The fuels formed millions of years ago from the remains of prehistoric plants and animals. Coal, oil and natural gas are fossil fuels.

Greenhouse Effect
The process by which gases like CO_2 in the atmosphere trap the Sun's heat and keep the planet warm enough for life to survive.

Global warming
The name given to the gradual warming up of the Earth's atmosphere, which is caused by burning too many fossil fuels.

Hydrocarbons
Substances that consist of hydrogen and carbon.

Pesticides
Chemicals used to control and kill pests, and protect crops.

Plankton
Tiny plants and animals that float in the sea.

Index

A
acid rain 5, 21, 31
alternative energy sources 26-27

C
coal 4, 6-9, 17, 18, 20, 21, 24, 25, 28, 30
coke 4, 18

E
efficient technology 25-5
electricity 4, 10, 16-17, 23, 25, 26, 27, 30

G
gases 5, 10-11, 15, 18, 25, 29, 30
global warming 5, 21, 31

H
hydrocarbons 10, 31

I
industrial uses 4, 5, 18, 19, 28

M
mining 8-9, 20

N
national grid 16

O
oil 4, 5, 10-15, 18, 19, 20, 21, 28, 29

P
peat 6, 7
petrochemicals 19
pollution 5, 21
power stations 4, 16-17

R
recycling 22-3

T
topping cycle 30

PRINTED IN BELGIUM BY proost INTERNATIONAL BOOK PRODUCTION